21 SUPER SIMPLE Chemistry EXPERIMENTS

Rebecca W. Keller, Ph.D.

Illustrations: Rebecca W. Keller, Ph.D.

Copyright © 2011 Gravitas Publications, Inc.

All rights reserved. No part of this publication may be reproduced, stored in a retrieval system, or transmitted, in any form or by any means, electronic, mechanical, photocopying, recording, or otherwise, without prior written permission from the publisher. However, this publication may be photocopied without permission from the publisher only if the copies are to be used for teaching purposes within a family.

21 Super Simple Chemistry Experiments

ISBN 13: 978-1-936114-37-5

Published by Gravitas Publications, Inc.
4116 Jackie Road SE, Suite 101
Rio Rancho, NM 87124
www.gravitaspublications.com

Printed in United States

What are Super Simple Science Experiments?

Super Simple Science Experiments are experiments that focus on one aspect of scientific investigation. Doing science requires students to develop different types of skills. These skills include the ability to make good observations, turning observations into questions and/or hypotheses, building and using models, analyzing data, using controls, and using different science tools including computers.

Super Simple Science Experiments break down the steps of scientific investigation so that students can focus on one aspect of scientific inquiry. The experiments are simple and easy to do yet they are *real* science experiments that help students develop the skills they need for *real* scientific investigations.

Each experiment is one page and lists a short objective, the materials needed, and a brief outline of the experiment, and includes any graphics or illustrations needed for the experiment. The skill being explored is listed in the upper right hand corner of each page. Any additional pages required are included at the back of the book.

Getting Started

Below is a list of the materials and equipment needed for all of the chemistry experiments in this book. You can collect all the materials ahead of time and place them in a storage bin or drawer.

Materials at a Glance

Foods

apple
baking soda
bread
celery
chewing gum
gumdrops
lemon juice
marshmallows, large
marshmallows, small
milk
peanut butter
potatoes, several
red cabbage, 1 head
salt
soda pop, "healthy"
soda pop, brown
soda pop, clear
soda, club
sugar, white table
vinegar, apple cider
vinegar, clear
water, carbonated
water, distilled
water, filtered
water, mineral
water, tap
water, well

Materials

Super Simple Science Experiments Laboratory Notebook
balloons
ballpoint ink pen, black
ballpoint ink pens, various colors
bean seedlings in soil (2)
borax
coffee filter paper, white
cotton cloth
glue, white
iodine, 2% tincture (sodium iodide)
isopropyl alcohol
liquid laundry starch
matches
pencil
pond water
popsicle stick
small candle
soap
string, white
tape
toothpicks

Equipment

container, mason jar
container, quart size
containers or jars, clear
cooking pot
freezer
knife
measuring cups
measuring spoons
mixing spoon
pie dishes, shallow (2)
scissors
stopwatch

Laboratory Safety

Most of these experiments use household items. However, some items, such as iodine, are extremely poisonous. Extra care should be taken while working with all chemicals in this series of experiments. The following are some general laboratory precautions that should be applied to the home laboratory:

Never put things in your mouth without explicit instructions to do so. This means that food items should not be eaten unless tasting or eating is part of the experiment.

Use safety glasses while working with glass objects or strong chemicals such as bleach.

Wash hands before and after handling chemicals.

Use adult supervision while working with iodine and while conducting any step requiring a stove or a knife.

Table of Contents

1.	Sodium	1
2.	Lead	2
3.	Model of glucose	3
4.	Single, double and triple bonds	4
5.	Methane	5
6.	Benzene	6
7.	Growing a sugar crystal	7
8.	Testing for acids	8
9.	Testing for bases	9
10.	Soda pop test	10
11.	Water test	11
12.	Combustion	12
13.	Separating ink	13
14.	Testing for starch	14
15.	Cold balloons	15
16.	Polymer cross-linking	16
17.	Polymer cross-linking with Borax	17
18.	Gumdrop model of DNA	18
19.	Acid rain	19
20.	Osmosis	20
21.	Everyday chemistry	21
	Periodic table of elements	22

1. Sodium

using resources

Objective

To become familiar with the periodic table of elements by learning about the element sodium

Materials

pencil
Super Simple Science Experiments
    Laboratory Notebook

Questions

❶ Look at the information for the element sodium shown in the illustration to the right. This is how the individual elements are often shown in a periodic table.

❷ The number of protons is found in the upper left hand corner. How many protons does sodium have? _____

❸ The atomic weight is found below the elemental name. What is the atomic weight? _____

❹ The symbol for sodium is found above the elemental name. What is the symbol for sodium? _____

❺ The number of electrons equals the number of protons. How many electrons does sodium have? _____

❻ The number of neutrons equals the atomic weight minus the number of protons (rounded to the nearest whole number). How many neutrons does sodium have? _____

Results and Conclusions

The periodic table of elements organizes chemical information about the elements. Knowing how to use the periodic table to quickly find the information it contains is an essential skill in chemistry. The symbol for an element often comes from the first letter of the elemental name. However, the symbol for sodium comes from the Latin word *natrium*.

2. Lead

using resources

Objective

To become familiar with the periodic table of elements by learning about the element lead

Materials

pencil
Super Simple Science Experiments
   Laboratory Notebook

Questions

① Look at the information for the element lead shown in the illustration to the right. This is how the individual elements are often shown in a periodic table.

② The number of protons is found in the upper left hand corner. How many protons does lead have? _____

③ The atomic weight is found below the elemental name. What is the atomic weight of lead? _____

④ The symbol for lead is found above the elemental name. What is the symbol for lead? _____

⑤ The number of electrons equals the number of protons. How many electrons does lead have? _____

⑥ The number of neutrons equals the atomic weight minus the number of protons (rounded to the nearest whole number). How many neutrons does lead have? _____

Results and Conclusions

The periodic table of elements organizes chemical information about the elements. Knowing how to use the periodic table to quickly find the information it contains is an essential skill in chemistry. The symbol for an element often comes from the first letter of the elemental name. However, the symbol for lead comes from the Latin word *plumbum*, which means "album" or "tin."

Chemistry 3

3. Model of glucose

building models

Objective

To study the chemical structure of glucose by building a model using marshmallows and toothpicks

Materials

pencil
large marshmallows
small marshmallows
toothpicks
Super Simple Science Experiments
   Laboratory Notebook

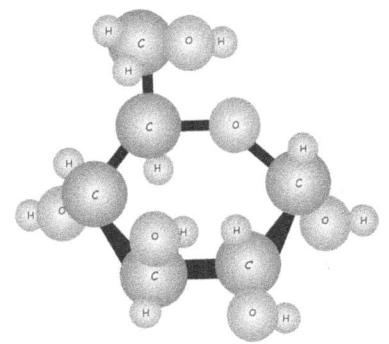

Experiment

① Glucose is a six-carbon sugar that forms a ring structure.
② Look at the image of glucose above. Observe the carbon, hydrogen, and oxygen atoms.
③ Using the large marshmallows for the carbon and oxygen atoms, the small marshmallows for the hydrogen atoms, and the toothpicks for bonds, create a marshmallow model for glucose.
④ Notice if the model is floppy or stiff, and whether or not you can bend it into a different shape.
⑤ Draw the shapes in your Laboratory Notebook.

Results and Conclusions

Using models is an important part of scientific investigation. Although models have limitations, they can help scientists understand possible shapes and behaviors of molecules.

Glucose is often drawn in textbooks as a planar circular molecule, but ringed structures, such as glucose, can take on different shapes such as "boat" or "chair" conformations.

circular

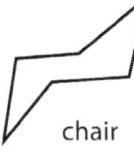
chair

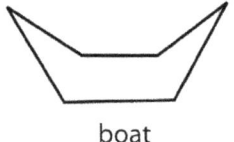
boat

4. Single, double, and triple bonds

building models

Objective

To use models to explore single, double, and triple bonds

Materials

pencil
large marshmallows
toothpicks
Super Simple Science Experiments
  Laboratory Notebook

single bond

double bond

triple bond

Experiment

❶ Atoms like carbon, nitrogen, and phosphorus can form single, double, and triple bonds with other atoms.

❷ Look at the image above that shows single, double, and triple bonds.

❸ Using marshmallows and toothpicks, create a marshmallow model for each type of bond: single, double, and triple.

❹ Notice how the marshmallows on each type of bond can or cannot rotate.

❺ Write your observations in your Laboratory Notebook.

Results and Conclusions

Using models is an important part of scientific investigation. Although models have limitations, they can help scientists understand possible shapes and behaviors of molecules.

Molecules are put together with single, double, or triple bonds or a combination of these. Single bonds allow the atoms to rotate around the bond axis. A single bond will result in a molecule that can be very flexible and floppy. Double and triple bonds restrict the rotation of atoms along the bond axis. As a result, molecules with double or triple bonds are more rigid and stiff.

5. Methane

building models

Objective

To explore a methane molecule using models

Materials

- pencil
- large marshmallows
- small marshmallows
- toothpicks
- Super Simple Science Experiments Laboratory Notebook

(methane)

Experiment

❶ Methane is made of one carbon atom and four hydrogen atoms.

❷ Look at the above drawing of a methane molecule.

❸ Using the large marshmallows for carbon, the small marshmallows for hydrogen, and the toothpicks for bonds, create a marshmallow model for methane.

❹ Make a planar molecule, with all of the marshmallows on the same plane. Then make another molecule that has the small marshmallows extending outward in all directions from the center large marshmallow, maximizing the distance between the hydrogen atoms.

❺ Observe any differences between the two shapes, and write your observations in your Laboratory Notebook.

Results and Conclusions

Methane forms a tetrahedral molecule rather than a planar molecule. The central carbon atom forms a single bond with each of the four hydrogen atoms. The hydrogen atoms extend from the central carbon in all directions so that the distance between each hydrogen atom is maximized.

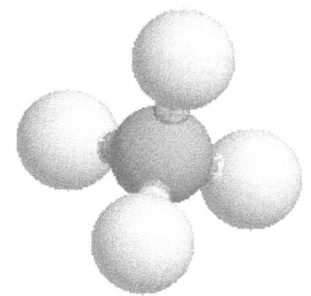

tetrahedral shape

6. Benzene

> building models

Objective

To use models to explore the molecule benzene

Materials

- pencil
- large marshmallows
- small marshmallows
- toothpicks
- Super Simple Science Experiments Laboratory Notebook

Benzene, C₆H₆

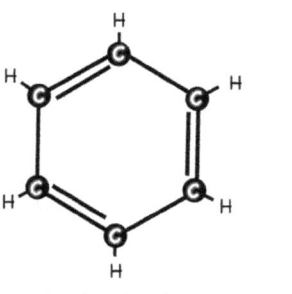
molecular structure showing atoms

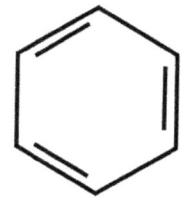
molecular structure without showing atoms

Experiment

❶ Benzene is a ringed structure made of six carbon atoms and six hydrogen atoms.

❷ Look at the drawings of benzene shown above.

❸ Using the large marshmallows for carbon, the small marshmallows for hydrogen, and the toothpicks for bonds, create a marshmallow model for benzene.

❹ Take your model and gently twist it in your hands without breaking any of the bonds. Make another model with single bonds replacing the double bonds, and notice if this model bends more easily.

❺ Observe the flexibility of your benzene model, and compare this to the ringed structure with no double bonds. Write your observations in your Laboratory Notebook.

Results and Conclusions

Benzene forms a planar ring-shaped structure with alternating double bonds. The double bonds make benzene a stiff molecule that does not change shape. Unlike ringed sugar molecules (Experiment 3), benzene does not form "chair" or "boat" conformations.

Chemistry 7

7. Growing a sugar crystal

observation

Objective

To make careful observations while growing a sugar crystal

Materials

- pencil
- popsicle stick
- white granulated table sugar
- water
- mason jar
- white string 6-12 inches in length
- Super Simple Science Experiments Laboratory Notebook
- measuring spoons
- measuring cup
- cooking pot
- mixing spoon

Experiment

1. Boil 2 cups of water. Remove the pan from the heat carefully to avoid getting splashed by the hot water.
2. Stir in the sugar very slowly, one tablespoonful at a time, swirling the solution after adding each tablespoonful.
3. Keep adding sugar until the sugar stops dissolving. You should start to see the sugar collect at the bottom of the pan.
4. Carefully pour the sugar water into the mason jar.
5. Attach the string to the popsicle stick, and adjust the length of the string so that it drops into the mason jar several inches deep.
6. Allow the jar to sit at room temperature, and observe how the sugar crystals grow. Record your observations in your Laboratory Notebook

Results and Conclusions

When sugar is added to hot water until it stops dissolving, a super-saturated solution is formed. As the water cools, the sugar molecules come out of solution, and if the cooling happens slowly enough, the sugar molecules will have enough time to form crystals.

8. Testing for acids

observation

Objective

To use an acid-base indicator to test for acids

Materials

- pencil
- one head of red cabbage
- distilled water
- Super Simple Science Experiments Laboratory Notebook
- cooking pot
- knife
- clear containers or jars
- measuring spoons

Experiment

1. Cut the red cabbage in quarters and place one quarter in a clean cooking pot. Add a few cups of distilled water, and boil the cabbage until the water turns deep purple.
2. Remove the cabbage from the water and let the water cool. The red cabbage juice can now be used as an acid-base indicator.
3. Test a variety of household liquids. Put each liquid in a clean clear container, and add one to two tablespoons of red cabbage juice indicator.
4. Record your observations in your Laboratory Notebook.

Suggested Liquids

- lemon juice
- tap water
- clear vinegar
- clear soda pop

Results and Conclusions

Red cabbage juice can be used as an acid-base indicator. When red cabbage juice is added to an acid, the mixture will turn a bright pink. The indicator changes color as a function of pH. When the pH is lower than 7, as it is for an acid, the red cabbage juice indicator turns pink.

{Note: Refrigerate remaining cabbage juice for use in Experiments 9, 10, and 11}

9. Testing for bases

observation

Objective

To use an acid-base indicator to test for bases

Materials

- pencil
- head of red cabbage
- distilled water
- Super Simple Science Experiments Laboratory Notebook
- cooking pot
- knife
- clear containers or jars
- measuring spoons

If you saved the red cabbage juice indicator from Experiment 8, please skip to Step 3.

Experiment

Suggested Liquids

- mineral water
- baking soda water
- club soda
- dilute ammonia

❶ Cut the red cabbage in quarters and place one quarter in a clean cooking pot. Add a few cups of distilled water and boil the cabbage until the water turns deep purple.

❷ Remove the cabbage from the water and let the water cool. The red cabbage juice can now be used as an acid-base indicator.

❸ Test a variety of household liquids. Put each liquid in a clean clear container, and add one to two tablespoons of red cabbage juice indicator.

❹ Record your observations in your Laboratory Notebook.

Results and Conclusions

Red cabbage juice can be used as an acid-base indicator. When red cabbage juice is added to a base, the mixture will turn a bright green. The indicator changes color as a function of pH. When the pH is greater than 7, as it is for a base, the red cabbage juice indicator turns green.

{Note: Refrigerate the remaining cabbage juice for use in Experiments 10 and 11}

10. Soda pop test

observation

Objective

To use an acid-base indicator to test the pH of different sodas

Materials

- pencil
- one head of red cabbage
- distilled water
- at least 3 different kinds of soda
- 3 clear containers
- Super Simple Science Experiments Laboratory Notebook
- cooking pot
- knife
- measuring spoons

If you saved the red cabbage juice indicator from Experiment 8 or 9, please skip to Step 3.

Suggested Sodas

- a clear soda
- a "brown" soda
- a "healthy" soda

Experiment

1. Cut the red cabbage in quarters and place one quarter in a clean cooking pot. Add a few cups of distilled water and boil the cabbage until the water turns deep purple.
2. Remove the cabbage from the water and let the water cool. The red cabbage juice can now be used as an acid-base indicator.
3. Pour each soda into a separate clear container. Add two tablespoons of the red cabbage juice indicator to each. {Note: For the "brown" soda, dilute 1:3 in distilled water so that the color change will be visible.}
4. Record your observations in your Laboratory Notebook.

Results and Conclusions

Most soda contains phosphoric acid and will typically turn red cabbage juice indicator pink. However, it is interesting to find out if various kinds of soda differ. Is "healthy" soda less or more acidic than clear soda? Is brown soda less or more acidic than clear soda?

{Note: Refrigerate the remaining cabbage juice for use in Experiment 11}

11. Water test

observation

Objective

To use an acid-base indicator to test the pH of different kinds of water

Materials

- pencil
- one head of red cabbage
- knife
- distilled water
- cooking pot
- several different kinds of water
- several clear containers or jars
- Super Simple Science Experiments Laboratory Notebook

Suggested types of water

- tap water
- distilled water
- mineral water
- pond water
- well water
- filtered water

If you saved the red cabbage juice indicator from Experiment 8, 9, or 10 please skip to Step 3.

Experiment

1. Cut the red cabbage in quarters and place one quarter in a clean cooking pot. Add distilled water and boil the cabbage until the water turns deep purple.
2. Remove the cabbage from the water and let the water cool. The red cabbage juice can now be used as an acid-base indicator.
3. Using clear containers, test the different types of water by adding red cabbage juice acid-base indicator to each.
4. Record your observations in your Laboratory Notebook.

Results and Conclusions

All water is not the same. Some water may be acidic, and other types of water may be basic. Distilled water is neutral and will not turn acid-base indicator pink or green. However, tap water, pond water, and well water can contain metal ions or added chlorine or fluorine that could change the pH.

12. Combustion

observation

Objective

To observe whether oxygen is required for a candle to burn

Materials

pencil
small candle
clear glass jar large enough to fit over the candle
matches
stopwatch
Super Simple Science Experiments Laboratory Notebook

Experiment

❶ Place the candle on a flat surface.
❷ Light the candle and observe how it burns.
Record your observations in your Laboratory Notebook.
❸ Place the clear glass jar over the candle.
❹ Observe what happens to the flame. Using the stopwatch, note how long it takes for the flame to extinguish.
❺ Record your observations in your Laboratory Notebook.

Results and Conclusions

Combustion requires oxygen. In the mid 1700s French chemist Antoine Lavoisier demonstrated that oxygen was required to burn combustible substances. Lavoisier's theory of combustion explained how oxygen in the air combined with metals (and other combustible substances) to form "oxides." When a candle burns, the paraffin wax is pulled up the wick, and oxygen combines with the wax to produce water and carbon dioxide.

13. Separating ink

using science techniques

Objective

To separate ink by performing a technique called paper chromatography

Materials

black ballpoint ink pen
optional: ballpoint pens in additional colors
scissors
clear glass jar
white coffee filter paper
isopropyl (rubbing) alcohol
tape
Super Simple Science Experiments
    Laboratory Notebook

Experiment

1. Pour several inches of rubbing alcohol into the glass jar.
2. Cut off both ends of the ink tube from the pen, and swirl one end in the alcohol until the alcohol becomes lightly colored
3. Cut a strip of coffee filter paper 2-3 inches long and 0.5 inches wide.
4. Tape the filter paper to the side of the glass jar so that one end is immersed in the alcohol and the other end is dry.
5. Allow the ink-alcohol mixture to sit overnight and migrate up the filter paper.
6. Record your observations in your Laboratory Notebook.

Results and Conclusions

Paper chromatography is a technique used to separate small molecules from each other. Many brands of black ink contain several colors mixed together to make black. In this experiment the different colors are separated by paper chromatography. Repeat the experiment using other colors of ink.

14. Testing for starch

using science techniques

Objective

To use iodine to test food for the presence of starch

Materials

2% tincture of iodine (sodium iodide)
a slice of each of the following:
- potato
- celery
- bread
- apple

Super Simple Science Experiments Laboratory Notebook

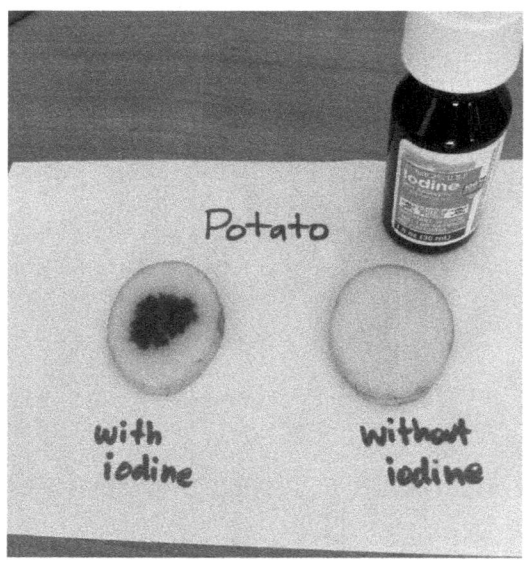

Experiment

NOTE: Iodine is poisonous! Do not eat. Wash hands after use.

1. Place the food slices on a table surface or piece of paper.
2. Carefully open the tincture of iodine and place a few drops on each piece of food.
3. Observe the initial color and then whether or not the color changes over time.
4. Record your observations in your Laboratory Notebook.

Results and Conclusions

Foods that contain starch will react with sodium iodide to produce a deep black-purple color. The iodine bonds to the coiled starch molecules, resulting in a color change. Observe what happens if you add sodium iodide to other food and non-food items. Do they have starch? How can you tell?

15. Cold balloons

using formulas

Objective

To observe the relationship between volume (V) and temperature (T)

Materials

- pencil
- 2-3 balloons
- freezer
- Super Simple Science Experiments Laboratory Notebook

Charles's Law

$V \propto T$

V = Volume
T = Temperature

Experiment

1. Inflate a balloon with air.
2. Place it in the freezer, and allow it to cool for several minutes.
3. Remove the balloon from the freezer.
4. Record your observations in your Laboratory Notebook.
5. Allow the balloon to warm up.
6. Record your observations in your Laboratory Notebook.

Results and Conclusions

In the early 19th century the relationship between the volume of a gas and its temperature was discovered by the French scientists J.A.C. Charles and J. L. Gay-Lussac. Charles's Law says that the volume of a gas is proportional to the temperature. The symbol $\propto$ means "proportional." In other words, as the temperature decreases the volume decreases, and as the temperature increases the volume increases. This is easily observed by watching a balloon contract when cooled in a freezer and then watching how it expands as it warms up to room temperature.

16. Polymer cross-linking

observation

Objective

To observe how polymer properties change when a polymer undergoes cross-linking

Materials

pencil
white glue (one or more brands)
liquid laundry starch
jar
measuring cup
Super Simple Science Experiments
   Laboratory Notebook

Experiment

❶ Pour a small amount of glue on your fingertips. Note the color, texture, smell, and feel of the glue. Record your observations in your Laboratory Notebook.

❷ Pour about 1/4 cup of glue into a jar.

❸ Add enough liquid laundry starch to cover the glue completely.

❹ Massage the glue/laundry starch mixture with your fingers. Observe how the glue changes over time.

❺ Record your observations in your Laboratory Notebook.

Results and Conclusions

Glue is a polymer, and when liquid laundry starch is added to glue, the glue forms cross-linked bonds with itself and the laundry starch. When these bonds form, the properties of the glue change. The glue changes from sticky to rubbery and from a liquid to a condensed mass. What happens if you use a different brand of glue? Do you get the same results?

Chemistry 17

17. Polymer cross-linking with borax

using comparisons

Objective

To observe how polymer properties change when a polymer undergoes cross-linking

Materials

pencil
white glue
borax
water
jar
measuring cup
measuring spoons
Super Simple Science Experiments
   Laboratory Notebook

Experiment

1. Pour a small amount of glue on your fingertips. Note the color, texture, smell, and feel of the glue. Record your observations in your Laboratory Notebook.
2. Pour about 1/4 cup of glue into a jar.
3. Mix 1 tablespoon of borax into 1 cup of water.
4. Add the borax water to the glue. Massage the mixture with your fingers. Observe how the glue changes over time.
5. Record your observations in your Laboratory Notebook.

Results and Conclusions

In Experiment 16 a glue and laundry starch mixture was created, and the properties of the glue were changed during the process of cross-linking. In this experiment the laundry starch was replaced with borax. Compare the two materials. Did both work equally well? Why or why not?

18. Gumdrop model of DNA

model building

Objective

To build a model of DNA

Materials

pencil
gumdrops
toothpicks
Super Simple Science Experiments
    Laboratory Notebook

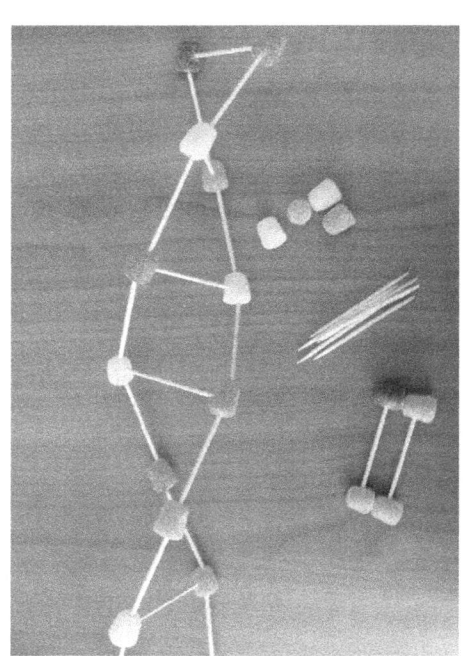

Experiment

❶ Create a "ladder" with the gumdrops and toothpicks.

❷ Twist the ends of the gumdrop ladder, until one side of the ladder wraps around the other side. This is a double helix.

❸ Record your observations in your Laboratory Notebook.

Results and Conclusions

DNA forms a double helix that has two sugar-phosphate backbone strands and bases connected along the center. A simple way to illustrate a DNA double helix is to create a gumdrop "ladder" and then twist the ladder at the ends to form the double helix. How many rungs are needed for the ladder to be able to twist? How far can the ladder twist? What does the model reveal about a DNA molecule?

19. Acid rain

using controls

Objective

To observe the effects of acid on a growing plant

Materials

pencil
apple cider vinegar
water
measuring cups
quart size container
2 bean seedlings planted in separate containers
Super Simple Science Experiments Laboratory Notebook

Experiment

1. Mark one bean plant "A" and the other "B."
2. Prepare acid water by adding 1/2 cup of apple cider vinegar to one quart of water. Label this water "A."
3. Put both bean seedlings in an area with plenty of sun. Water the bean seedling marked "A" with the acid water and the bean seedling marked "B" with regular water.
4. Observe the two seedlings for a few days, and record your observations in your Laboratory Notebook. Note any differences between plant "A" and plant "B."

Results and Conclusions

Acid rain slows the growth of plants and causes leaves to wilt and shrivel. In this experiment the apple cider water is acidic and can be used to simulate the effects of acid rain.

In this experiment a "control" (plant "B") was compared to the test plant (plant "A"). Controls are important in science because they allow the researcher to change one variable (the pH of the water) while keeping all other variables the same. In this experiment two identical types of plants (bean plants) were grown under similar conditions (sun, soil, time, etc.) except for the pH of the water. The effects of the acid water can be directly observed by comparing the test plant "A" to the control plant "B."

20. Osmosis

observation

Objective

To observe the effects of high salt concentration on a potato

Materials

pencil
a potato sliced lengthwise into several pieces
knife
water
salt
2 shallow pie dishes
measuring cup
measuring spoons
Super Simple Science Experiments Laboratory Notebook

Experiment

1. Pour a cup of water into each pie dish.
2. To one pie dish add two tablespoons of salt and stir.
3. Add several pieces of sliced potato to each pie dish and allow to sit for 15 minutes.
4. Observe the potato slices. Note any differences, and record you observations in your Laboratory Notebook.

Results and Conclusions

Living things are made of water. When salt is added to water, the number of salt molecules per unit volume increases. (The number of molecules per unit volume is called the concentration of a solution.) Salt water has a higher concentration of salt molecules than regular water. In this experiment water moves from areas of low salt concentration to areas of high salt concentration during the process of osmosis. Osmosis can be observed as the water in the potato moves out of the potato into the salt water solution in the dish, causing the potato to become limp.

21. Everyday Chemistry

problem solving

Objective

To test different types of cleaners

Materials

pencil
chewing gum
peanut butter
soapy water
cotton cloth
Super Simple Science Experiments
  Laboratory Notebook

Experiment

❶ Chew the gum until soft.
❷ Place the chewed gum on the cotton cloth and massage until stuck.
❸ Attempt to wash the cotton cloth with the soapy water. Record your observations in your Laboratory Notebook
❹ Massage the peanut butter into the gum and cloth. Record your observations in your Laboratory Notebook.

Results and Conclusions

Chemistry can be useful for everyday events. Getting gum stuck on a piece of clothing or in hair happens often. How do you choose what to use to remove the gum? Using soapy water is not as effective as using an oil-based product like peanut butter, because gum is a polymer and will dissolve more easily in an oil-based product than in soapy water.

The Periodic Table of Elements

http://www.gravitaspublications.com

*The lanthinide series: Elements 58-71

**The actinide series: Elements 90-103

www.ingramcontent.com/pod-product-compliance
Lightning Source LLC
Chambersburg PA
CBHW081357040426
42451CB00017B/3482